Pinpoint Math™

Student Booklet
Level E

Volume 3
Add and Subtract

The McGraw·Hill Companies

Photo Credits

©iStock International Inc., cover.

Acknowledgements

Content Consultant:

Linda Proudfit, Ph.D.

After earning a B.A. and M.A in Mathematics from the University of Northern Iowa, Linda Proudfit taught junior- and senior-high mathematics in Iowa. Following this, she earned a Ph.D. in Mathematics Education from Indiana University. She currently is Coordinator of Elementary Education and Professor of Mathematics Education at Governors State University in University Park, IL.

Dr. Proudfit has made numerous presentations at professional meetings at the local, state, and national levels. Her main research interests are problem solving and algebraic thinking.

www.WrightGroup.com

Printed in USA.

Send all inquiries to:
Wright Group/McGraw-Hill
P.O. Box 812960
Chicago, IL 60681

ISBN 978-1-40-4568020
MHID 1-40-4568026

2 3 4 5 6 7 8 9 10 RHR 13 12 11 10 09

Contents

Objectives

Volume 3: Add and Subtract

Topic 7 Add or Subtract 1- and 2-Digit Numbers

Lesson	Objective	Pages
Topic 7 Introduction	**7.1** Solve problems involving one-digit numbers added to two-digit numbers. **7.2** Add two 2-digit numbers with and without regrouping. **7.3** Subtract two 2-digit numbers with and without regrouping.	1
Lesson 7-1 1-Digit and 2-Digit Numbers	**7.1** Solve addition problems with one- and two-digit numbers.	2–4
Lesson 7-2 Add 2-digit Numbers	**7.2** Add two 2-digit numbers with and without regrouping.	5–7
Lesson 7-3 Subtract 2-Digit Numbers	**7.3** Subtract two 2-digit numbers with and without regrouping.	8–10
Topic 7 Summary	Review skills involving adding and subtracting 1- and 2-digit numbers.	11
Mixed Review 7	Maintain concepts and skills.	12

Topic 8 Add or Subtract Multidigit Numbers

Lesson	Objective	Pages
Topic 8 Introduction	**8.1** Use mental arithmetic to find the sum or difference of two two-digit numbers. **8.2** Estimate sums and differences. **8.3** Find the sum or difference of two whole numbers up to three digits long. **8.4** Find the sum or difference of two whole numbers between 1 and 10,000.	13
Lesson 8-1 Add or Subtract Mentally	**8.1** Use mental arithmetic to find the sum or difference of two two-digit numbers.	14–16
Lesson 8-2 Add and Subtract with Estimation	**8.2** Estimate sums and differences of 2-digit numbers.	17–19
Lesson 8-3 Add and Subtract 3-Digit Numbers	**8.3** Find the sum or difference of two whole numbers up to three digits long.	20–22
Lesson 8-4 Whole Numbers to 10,000	**8.4** Find the sum or difference of two whole numbers between 0 and 10,000.	23–25
Lesson 8-5 Add and Subtract Multidigit Numbers	**8.5** Demonstrate an understanding of, and the ability to use, standard algorithms for the addition and subtraction of multidigit numbers.	26–28
Topic 8 Summary	Review addition and subtraction of multidigit numbers.	29
Mixed Review 8	Maintain concepts and skills.	30

Tutorial Guide

Each of the standards listed below has at least one animated tutorial for students to use with the lesson that matches the objective. If you are using the electronic components of *Pinpoint Math,* you will find a complete listing of Tutorial codes and titles when you access them either online or via CD-ROM.

Level E

Standards by topic	**Tutorial codes**
Volume 3 Add or Subtract	
Topic 7 Add or Subtract 1- and 2-Digit Numbers	
7.1 Solve addition problems with one- and two-digit numbers.	7a Using Models to Solve Word Problems
7.2 Add two 2-digit numbers with and without regrouping.	7b Using the Partial-Sums Algorithm to Add
7.2 Add two 2-digit numbers with and without regrouping.	7c Solving Word Problems, Example A
7.2 Add two 2-digit numbers with and without regrouping.	7a Using Models to Solve Word Problems
7.3 Subtract two 2-digit numbers with and without regrouping.	7d Using the Same-Change Rule to Subtract
Topic 8 Add or Subtract Multidigit Numbers	
8.2 Estimate sums and differences of 2-digit numbers.	8a Using the Partial Sums Algorithm to Add
8.2 Estimate sums and differences of 2-digit numbers.	8b Using the Same Change Rule to Subtract
8.3 Find the sum or difference of two whole numbers up to three digits long.	8c Using the Standard Addition Algorithm, Example A
8.3 Find the sum or difference of two whole numbers up to three digits long.	8d Using the Standard Subtraction Algorithm, Example A
8.4 Find the sum or difference of two whole numbers between 0 and 10,000.	8e Using the Standard Addition Algorithm, Example B
8.4 Find the sum or difference of two whole numbers between 0 and 10,000.	8f Using the Standard Subtraction Algorithm, Example B
8.5 Demonstrate an understanding of, and the ability to use, standard algorithms for the addition and subtraction of multidigit numbers.	8c Using the Standard Addition Algorithm, Example A
8.5 Demonstrate an understanding of, and the ability to use, standard algorithms for the addition and subtraction of multidigit numbers.	8d Using the Standard Subtraction Algorithm, Example A
8.5 Demonstrate an understanding of, and the ability to use, standard algorithms for the addition and subtraction of multidigit numbers.	8e Using the Standard Addition Algorithm, Example B
8.5 Demonstrate an understanding of, and the ability to use, standard algorithms for the addition and subtraction of multidigit numbers.	8f Using the Standard Subtraction Algorithm, Example B

Topic 7

Add or Subtract 1- and 2-Digit Numbers

Topic Introduction

Complete with teacher help if needed.

1. Subtract.

a. 76
− 43

b. 62
− 35

Objective 7.3: Subtract two 2-digit numbers with and without regrouping.

2. What is the sum for the addition problem shown with base ten blocks?

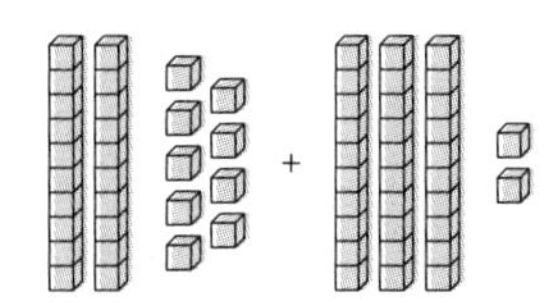

a. Write the addition problem using numbers. ______

b. There are 5 tens and ______ ones.

c. Trade 10 ones for ______ ten.

d. The sum is ______.

Objective 7.2: Add two 2-digit numbers with and without regrouping.

3. Subtract: 42 − 19.

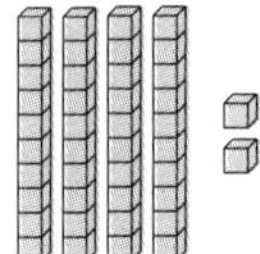

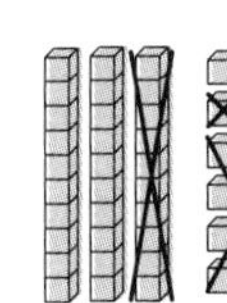

a. Trade 1 ten for ______ ones.

b. Cross out 1 ten and ______ ones.

c. ______ tens ______ ones are left.

d. The difference is ______.

Objective 7.3: Subtract two 2-digit numbers with and without regrouping.

4. Add.

a. 24
+ 5

b. 36
+ 8

Objective 7.1: Solve problems involving one-digit numbers added to two-digit numbers.

Lesson 7-1 1-Digit and 2-Digit Numbers

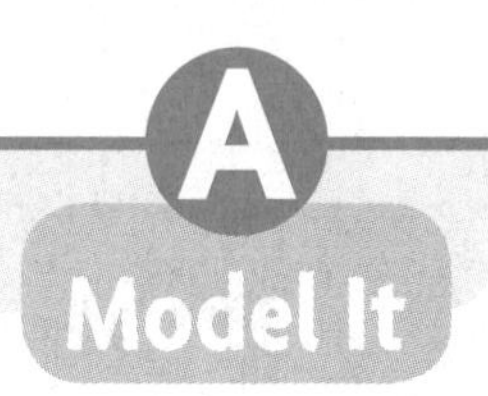

Activity 1

Use ones and tens blocks to find 23 + 9.

Model each number. Combine the ones.
Trade 10 ones for 1 ten.

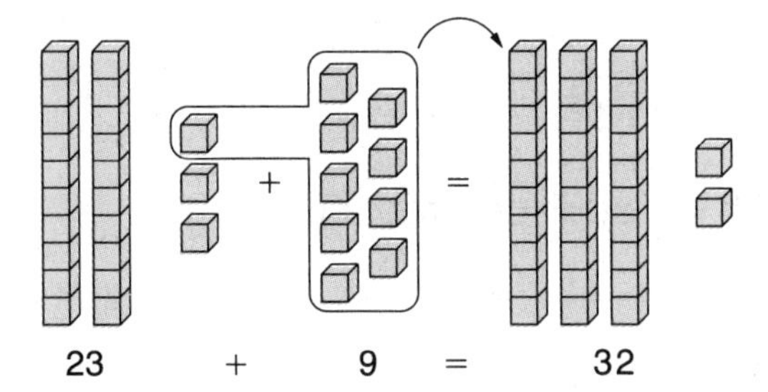

Practice 1

Use ones and tens blocks to find 16 + 5.

How many ones in all? ______

After you trade 10 ones for 1 ten, there are ______ tens and ______ one in all.

16 + 5 = ______

Activity 2

Use ones and tens blocks to find 45 + 7.

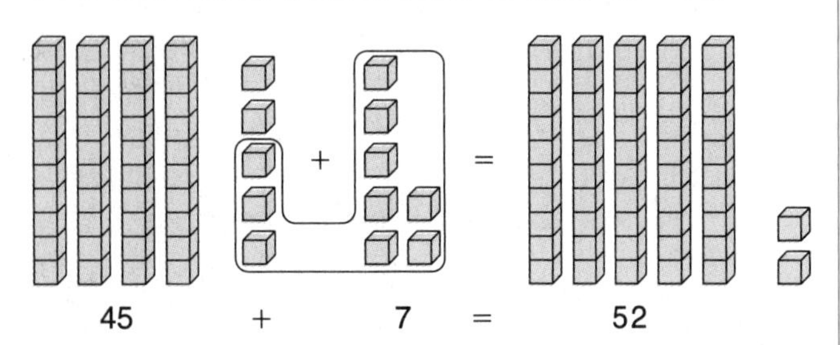

Add ones.
Trade 10 ones for 1 ten.
Add tens.

$$\begin{array}{r} ^{1} \\ 45 \\ +\ 7 \\ \hline 52 \end{array}$$

42 + 7 = 52

Practice 2

Use ones and tens blocks to find 58 + 7.

Add ones.

$$\begin{array}{r} 58 \\ +\ 7 \\ \hline \end{array}$$

Trade 10 ones for ______ ten.

Add tens.

On Your Own

Lisa collected 24 stuffed animals for charity. She wants to collect 8 more. How many will she have then?

She will have ______ stuffed animals.

Write About It

When you add, sometimes you need to trade 10 ones for a ten and sometimes you don't. When do you need to trade?

Objective 7.1: Solve addition problems with one- and two-digit numbers.

Lesson 7-1 1-Digit and 2-Digit Numbers

B Understand It

Example 1

Use the number line to help you add.
37 + 6 = ?

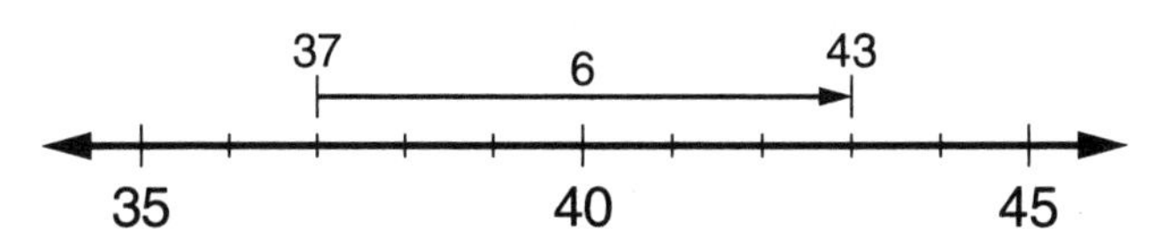

Start at 37.
Move 6 units to the right.
You stop at 43.

37 + 6 = 43

Practice 1

Use the number line to help you add.
29 + 5 = ?

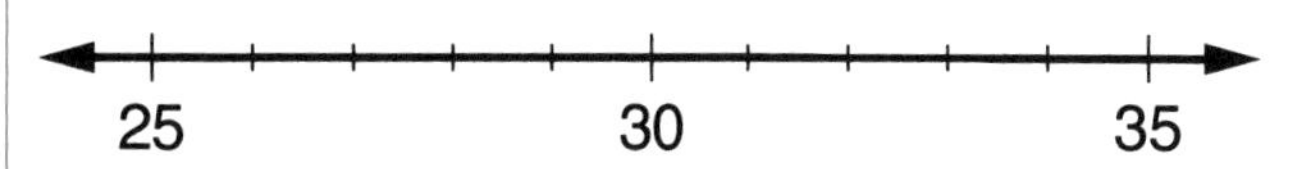

Start at ______.

Move ______ units to the right.

You stop at ______.

29 + 5 = ______

Example 2

Corey has 23 paper clips. A friend gives him 9. How many paper clips does Corey have now?

Add to find how many in all.

Add ones. Trade ones. Add tens.

$$\begin{array}{r} 23 \\ 9 \\ \hline \end{array} \qquad \begin{array}{r} \scriptstyle 1 \\ 23 \\ +\ 9 \\ \hline 2 \end{array} \qquad \begin{array}{r} \scriptstyle 1 \\ 23 \\ +\ 9 \\ \hline 32 \end{array}$$

Corey has 32 paper clips.

Practice 2

Lourdes had 68 marbles. She wins 3 more. How many does she have now?

Lourdes has ______ marbles.

On Your Own

Brittney buys 46 stamps. She already had 7 stamps. How many stamps does she have now?

Brittney has ______ stamps.

Write About It

Describe how to show the regrouping in 87 + 4.

Objective 7.1: Solve addition problems with one- and two-digit numbers.

1-Digit and 2-Digit Numbers

1. Write the addition problem for the picture. Solve.

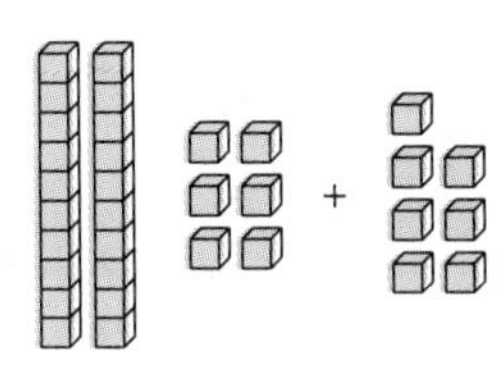

______ + ______ = ______

2. Use a number line to solve.

a. 36 + 2 = ______ **b.** 76 + 5 = ______

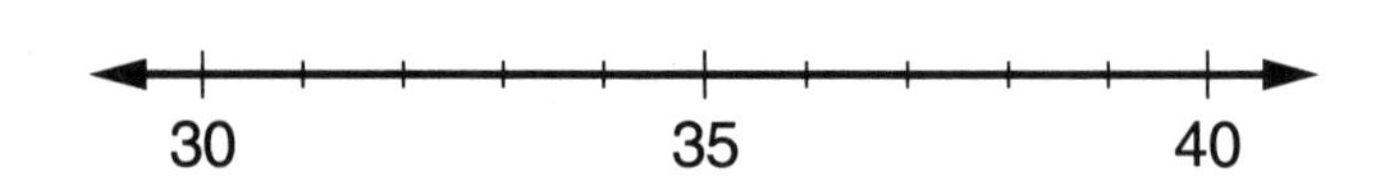

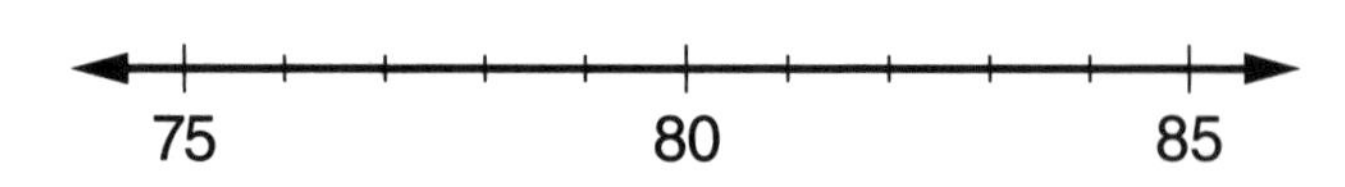

3. Add.

a. $92 + 3$

b. $53 + 8$

c. $63 + 7$

d. $72 + 9$

4. Tran spends 47 minutes on his art project and 9 minutes writing a poem. How long did he work on these projects?

A 56 minutes **B** 46 minutes

C 40 minutes **D** 50 minutes

5. Allana says 43 + 8 is 41. Is her answer correct? Explain why or why not.

6. To find 54 + 6, is it necessary to trade? Explain.

7. Mike had some magazines. He gives 5 to a friend. He has 29 left. How many did he have to start with?

____

8. 68 books are on the classroom bookshelves. 9 books are checked out. How many books belong to the classroom library?

____

Objective 7.1: Solve addition problems with one- and two-digit numbers.

Lesson 7-2 Add 2-Digit Numbers

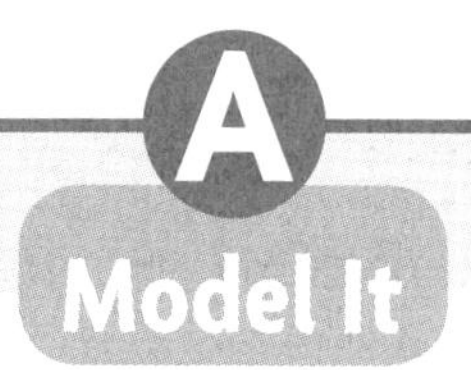

Words to Know **Regroup** means to exchange amounts of equal value to rename a number.

Activity 1

Use base ten blocks to find 38 + 25.

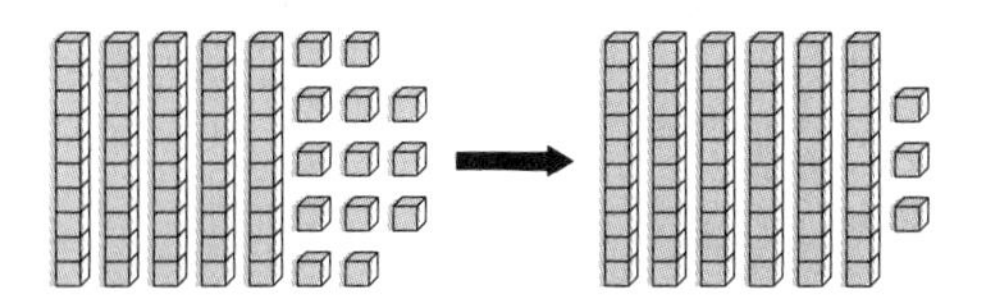

There are 13 ones in the total.

Regroup as 1 ten and 3 ones.

38 + 25 = 63

Practice 1

Use base ten blocks to find 27 + 24.

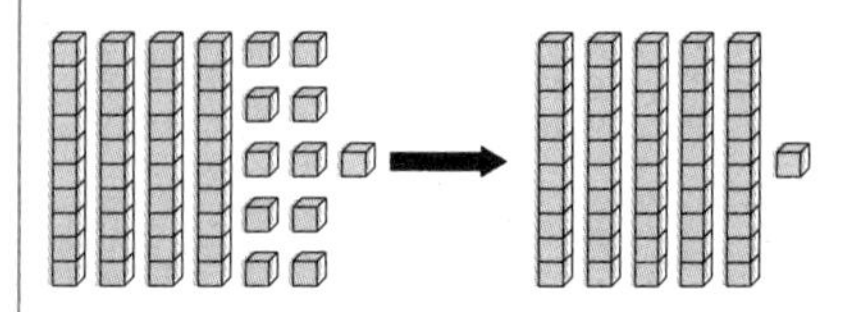

There are ______ ones.

Regroup as ______ ten(s) and ______ one(s).

27 + 24 = ______

Activity 2

Use base ten blocks to find 18 + 96.

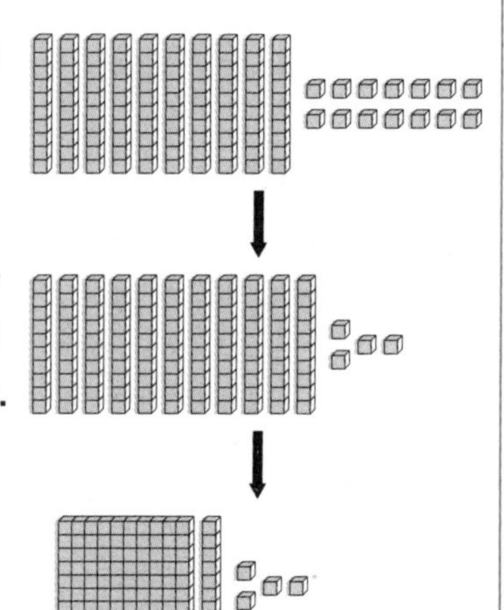

Regroup the ones as 1 ten and 4 ones.

Regroup the tens as 1 hundred and 1 ten.

18 + 96 = 114

Practice 2

Use base ten blocks to find 79 + 62.

First there are ______ ones. Regroup as ______ ten and ______ one. That makes ______ tens. Regroup as ______ hundred and ______ tens.

79 + 62 = ______

On Your Own

Regroup these base ten blocks.

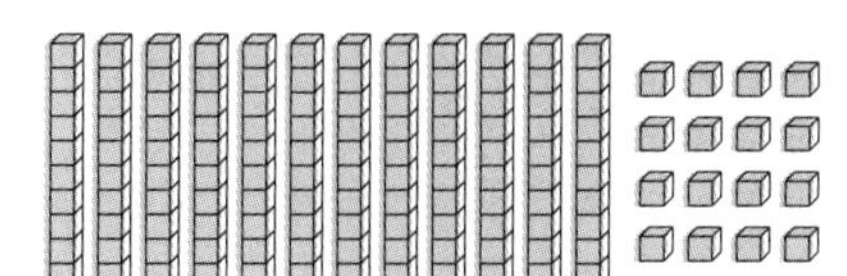

Write About It

Find the sum. What base ten blocks represent the answer?

$$\begin{array}{r} 77 \\ +\ 38 \\ \hline \end{array}$$

Objective 7.2: Add two 2-digit numbers with and without regrouping.

Lesson 7-2 Add 2-Digit Numbers

Example 1

Find the sum.

$$\begin{array}{r} \overset{1}{5}6 \\ +\ 18 \\ \hline 74 \end{array}$$

There are 14 ones. Regroup as 1 ten and 4 ones. Write the 1 ten as a small 1 in the tens column.

Practice 1

Find the sum.

$$\begin{array}{r} 76 \\ +\ 17 \\ \hline \_\_\_\_ \end{array}$$

There are ______ one(s).

Regroup as ______ ten(s)

and ______ one(s).

Example 2

Find each sum. Circle the problem that needs regrouping.

$$\begin{array}{r} 32 \\ +\ 65 \\ \hline 97 \end{array} \qquad \begin{array}{r} \overset{1}{1}8 \\ +\ 45 \\ \hline 63 \end{array}$$

13 ones need regrouping as 1 ten, 3 ones.

Practice 2

Find each sum. Circle the problem that needs regrouping.

$$\begin{array}{r} 59 \\ +\ 12 \\ \hline \_\_\_\_ \end{array} \qquad \begin{array}{r} 46 \\ +\ 13 \\ \hline \_\_\_\_ \end{array}$$

______ ones need regrouping as ______ ten(s),

______ one(s).

On Your Own

Find each sum.

$$\begin{array}{r} 67 \\ +\ 43 \\ \hline \_\_\_\_ \end{array} \qquad \begin{array}{r} 89 \\ +\ 77 \\ \hline \_\_\_\_ \end{array}$$

Write About It

Does 47 + 16 need regrouping? Explain.

Objective 7.2: Add two 2-digit numbers with and without regrouping.

Add 2-Digit Numbers

1. What sum do these base ten blocks represent?

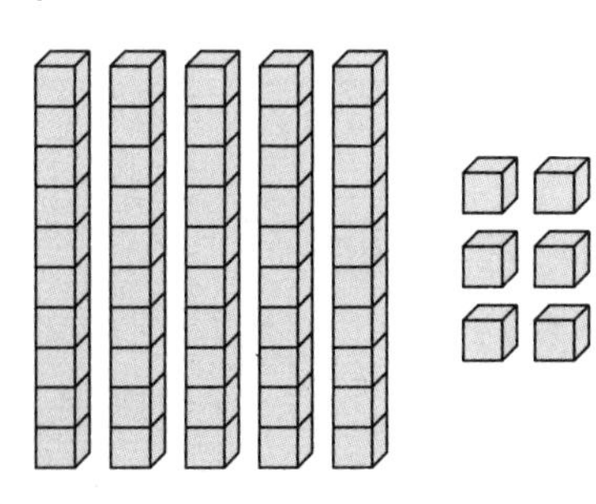

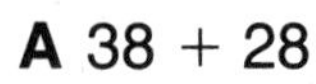

A 38 + 28 **B** 38 + 18

C 25 + 39 **D** 29 + 35

2. What is the missing digit?

$$\begin{array}{r} 42 \\ +\ 28 \\ \hline \end{array}$$

A 1 **B** 2

C 6 **D** 7

3. Students have the choice to take either art or drama. There are 24 students in drama. There are 57 students in art. How many students are in drama and art all together? Show your work.

There are ______ students in drama and art all together.

4. Alicia counts 28 swimmers in a large pool and 17 swimmers in a small pool. How many swimmers are in the pools all together? Show your work.

There are ______ swimmers in the pools all together.

5. Find each sum. Circle the problem that needs regrouping.

$$\begin{array}{r} 98 \\ +\ 29 \\ \hline \end{array} \qquad \begin{array}{r} 10 \\ +\ 82 \\ \hline \end{array}$$

6. Find each sum. Circle the problem that needs regrouping.

$$\begin{array}{r} 13 \\ +\ 25 \\ \hline \end{array} \qquad \begin{array}{r} 44 \\ +\ 85 \\ \hline \end{array}$$

7. Which of the following expressions needs regrouping?

A 15 + 14 **B** 27 + 43

C 41 + 51 **D** 12 + 74

8. Eric is adding 23 and 64. Does he need to regroup? Why or why not?

Objective 7.2: Add two 2-digit numbers with and without regrouping.

Lesson 7-3 Subtract 2-Digit Numbers

Words to Know **Regrouping** is exchanging amounts of equal value to rename a number. You can regroup 10 ones as 1 ten. You can regroup 1 ten as 10 ones.

Activity 1

Use base ten blocks to solve 42 – 5.

Think: Can I take away 5 ones from 2 ones? No.

Regroup to make 3 tens and 12 ones.

Now you can subtract 5.

42 – 5 = 37

Practice 1

Use base ten blocks to solve 26 – 9.

Think: Can I take away 9 ones from 6 ones?

______

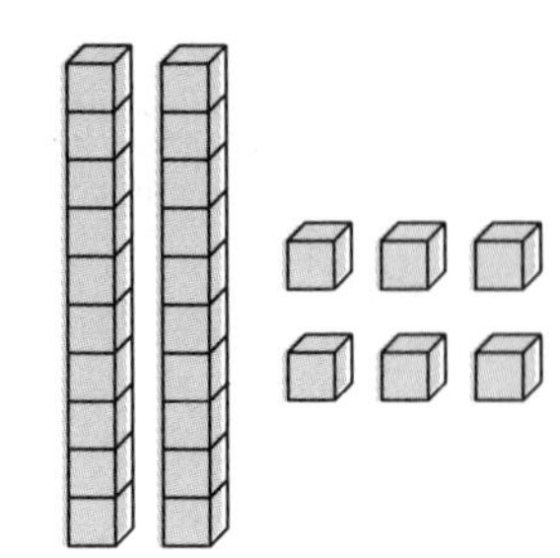

Regroup to make ______ ten and ______ ones.

Now you can subtract 9.
26 – 9 = 17

Activity 2

Find 94 – 38.
You can't subtract 8 ones from 4 ones.
Regroup to make 0 tens and 14 ones.
Subtract 8 ones.
Then subtract 3 tens from 8 tens.

94 – 38 = 56

Practice 2

Find 82 – 46.
You can't subtract 6 ones from 2 ones.
Regroup to make ______ tens and ______ ones.
Subtract ______ ones.
Then subtract ______ tens from ______ tens.
82 – 46 = ______

On Your Own

Find each difference. Show your work.

$$\begin{array}{r} 35 \\ -\ 6 \\ \hline \end{array} \qquad \begin{array}{r} 51 \\ -\ 8 \\ \hline \end{array}$$

You have 3 tens and 2 ones. You want to subtract 6 ones. Explain how to regroup.

Objective 7.3: Subtract two 2-digit numbers with and without regrouping.

Lesson 7-3 Subtract 2-Digit Numbers

B Understand It

Example 1

Subtract.

```
 7 14
 8̸4̸
 − 9
 ---
  75
```

Regroup 1 ten as 10 ones. Cross out the 8 and mark 7 above it to show how many tens we have. Then cross out the 4 and mark a small 14 above it to show how many ones we have. Then subtract.

Practice 1

Subtract.

```
  61
 − 7
 ---
 ___
```

Regroup. Now there are ______ tens and ______ ones.

Then subtract.

Example 2

Find the difference.

```
 0 13
 1̸3̸8
 − 52
 ----
   86
```

Regroup 1 hundred as 10 tens. Now there are 13 tens. Then subtract 5 tens and 2 ones.

Practice 2

Find the difference.

```
 108
 − 47
 ----
 ____
```

Regroup. Now there are ______ tens.

Then subtract ______ tens and ______ ones.

On Your Own

Find each difference. Show your work.

```
  40        74
 − 19      − 65
 ----      ----
 ____      ____
```

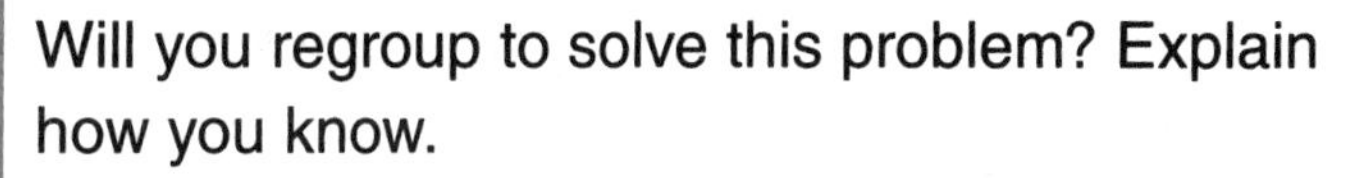

Write About It

Will you regroup to solve this problem? Explain how you know.

```
  52
 − 18
 ----
```

Objective 7.3: Subtract two 2-digit numbers with and without regrouping.

Subtract 2-Digit Numbers

1. Find each difference. Show your work.

   a. $\begin{array}{r} 82 \\ -\ 6 \\ \hline \end{array}$

   b. $\begin{array}{r} 67 \\ -\ 9 \\ \hline \end{array}$

2. Find each difference. Show your work.

   a. $\begin{array}{r} 73 \\ -\ 42 \\ \hline \end{array}$

   b. $\begin{array}{r} 70 \\ -\ 18 \\ \hline \end{array}$

3. Find each difference. Show your work.

   a. $\begin{array}{r} 137 \\ -\ 58 \\ \hline \end{array}$

   b. $\begin{array}{r} 184 \\ -\ 88 \\ \hline \end{array}$

4. Avery needs to regroup the blocks shown below. He wants to exchange a ten for ones. After regrouping, what blocks will Avery have? Circle the letter of the correct answer.

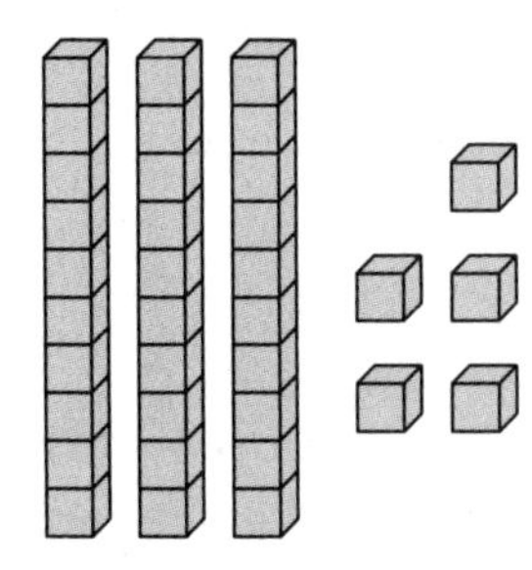

   A 3 tens, 15 ones **B** 2 tens, 15 ones

   C 3 tens, 10 ones **D** 2 tens, 10 ones

5. At a water park, 53 people are on the slides and 17 people are in the wave pool. How many more people are on the slides than in the wave pool? Show your work.

   $\begin{array}{r} 53 \\ -\ 17 \\ \hline \end{array}$

   There are ______ more people on the slides than in the wave pool.

6. What is the missing number? Explain.

   $\begin{array}{r} 87 \\ -\ 48 \\ \hline ?9 \end{array}$

7. Circle the letter of the problem in which you would regroup.

   A $\begin{array}{r} 85 \\ -\ 36 \\ \hline \end{array}$ **B** $\begin{array}{r} 59 \\ -\ 18 \\ \hline \end{array}$

   C $\begin{array}{r} 129 \\ -\ 14 \\ \hline \end{array}$ **D** $\begin{array}{r} 67 \\ -\ 50 \\ \hline \end{array}$

8. A class had 25 apples for a class party. If 17 of the apples were eaten, how many apples are left? Show your work.

   $\begin{array}{r} 25 \\ -\ 17 \\ \hline \end{array}$ There are ______ apples left.

Objective 7.3: Subtract two 2-digit numbers with and without regrouping.

Topic 7

Add or Subtract 1- and 2-Digit Numbers

Topic Summary

Choose the correct answer. Explain how you decided.

1. Tiffany has a collection of postcards from family vacations. If she collects 15 more postcards on the next vacation she will have a total of 53 postcards. How many postcards are in her collection now?

A 68

B 42

C 48

D 38

____

____

____

____

2. At a school assembly, there were 46 seventh graders and 35 sixth graders in attendance. How many students were at the assembly?

A 90

B 11

C 71

D 81

____

____

____

____

Objective: Review adding and subtracting 1- and 2-digit numbers.

Topic 7 Add or Subtract 1- and 2-Digit Numbers

Mixed Review

1. Find the sum.

a. $16 + 23 =$ ______

b. $79 + 11 =$ ______

c. $34 + 19 =$ ______

Volume 3, Lesson 7-2

2. Write 439 in word form.

____

Volume 1, Lesson 2-2

3. Multiply.

a. $7 \times 0 =$ ______

b. $1 \times 42 =$ ______

c. $0 \times 13 =$ ______

d. $18 \times 1 =$ ______

Volume 2, Lesson 5-3

4. Solve.

$65 - 48 =$ ______

Volume 3, Lesson 7-3

5. Add.

a. $4 + 2 + 6 =$ ______

b. $3 + 1 + 5 =$ ______

c. $8 + 2 + 7 =$ ______

Volume 2, Lesson 4-4

6. Which of the following is a fact in the family 4, 5, and 9? Circle the letter of the correct answer.

A $5 \times 4 = 20$

B $9 - 5 = 4$

C $9 \times 4 = 36$

D $9 + 5 = 14$

Volume 2, Lesson 4-7

Objective: Maintain concepts and skills.

Topic 8 Add or Subtract Multidigit Numbers

Topic Introduction

Complete with teacher help if needed.

1. Use mental math to add 20 + 50.

a. How many tens are in 20? ______

b. How many tens are in 50?______

c. Add the tens. ________________ tens

d. ______ tens = ______

Objective 8.1: Use mental arithmetic to find the sum or difference of two two-digit numbers.

2. Subtract: 1,392 – 417.

a. Set up the subtraction vertically.

$$\begin{array}{r} 1,392 \\ -\ 417 \\ \hline \end{array}$$

b. Subtract, regrouping as necessary.

$$\begin{array}{r} 1,392 \\ -\ 417 \\ \hline \end{array}$$

Objective 8.4: Find the sum or difference of two whole numbers between 1 and 10,000.

3. Estimate 42 – 19.

a. 42 rounds to ______.

b. 19 rounds to ______.

c. ______ – ______ = ______

d. The difference 42 – 19 is approximately ______.

Objective 8.2: Estimate sums and differences.

4. Add: 247 + 65.

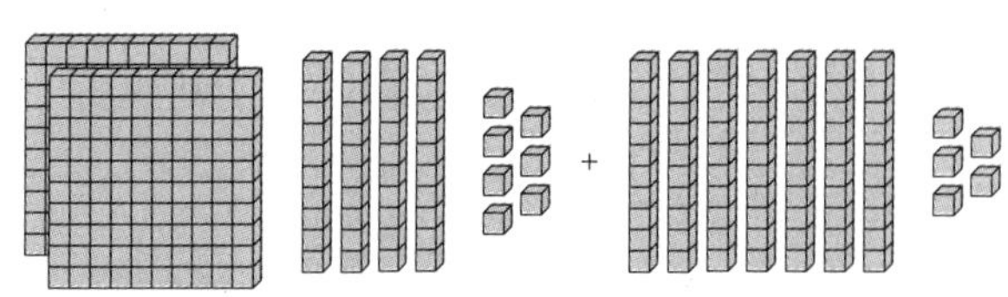

a. How many of each type of base ten blocks?

______ ones

______ tens

______ hundreds

b. Rewrite the number, trading when you can.

______ hundreds ______ tens ______ ones

Objective 8.3: Find the sum or difference of two whole numbers up to three digits long.

Lesson 8-1 Add or Subtract Mentally

A Model It

Activity 1

Find the number that is ten more than 34.

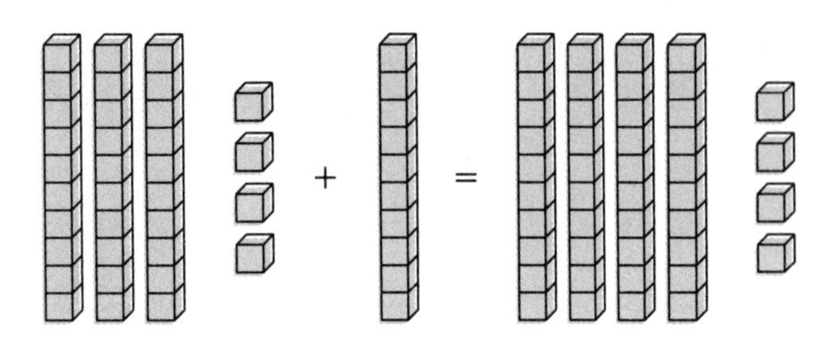

To find 10 more, add 1 more ten.

34 + 10 = 44

Ten more than 34 is 44.

Practice 1

Find the number that is 20 less than 56.

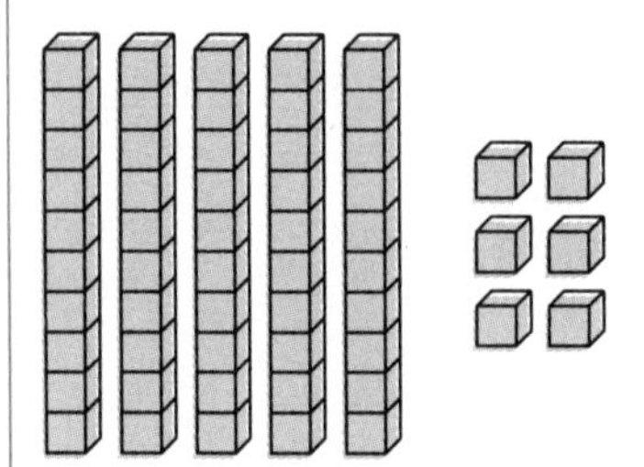

To find 20 less, subtract ______ tens.

56 − 20 = 36

Twenty less than 56 is ______.

Activity 2

Mentally, find the sum of 32 + 38.
Break apart and make 10 to use mental math.

32 + 38

Break apart: 32 + 30 + 8

Make 10: 32 + 8 + 30

40 + 30 = 70

The sum of 32 + 38 is 70.

Practice 2

Find the sum of 26 + 54 mentally.

Break apart and make 10 to use mental math.

26 + 54

Break apart: 26 + 50 + ____

Make 10: 26 + 4 + ____

____ + ____ = ____

26 + 54 = ______

On Your Own

Mentally, find the sum of 52 + 28.

52 + 28 = ______

Write About It

Explain how to find the sum of 19 + 31 using mental math.

Objective 8.1: Use mental arithmetic to find the sum or difference of two two-digit numbers.

Lesson 8-1 Add or Subtract Mentally

B

Understand It

Example 1

Find the difference of 60 – 25 mentally.

Break apart each number to subtract mentally.

Break apart: 60 = 50 + 10
25 = 20 + 5

Subtract:

$$\begin{array}{r} 50 \\ -20 \\ \hline 30 \end{array} \quad \begin{array}{r} 10 \\ -5 \\ \hline 5 \end{array}$$

Add the parts of the answer. 30 + 5 = 35

The difference of 60 – 25 is 35.

Practice 1

Find the difference of 70 – 15 mentally.

Break apart: 70 = ______ + 10

15 = ______ + ______

Subtract:

Add the parts. ______ + ______ = ______

The difference of 70 – 15 is ______.

Example 2

Subtract 94 – 71 mentally.

Break apart the second number to subtract mentally.

Break apart: 71 = 70 + 1

Subtract: 94 – 70 = 24
24 – 1 = 23

94 – 71 = 23

Practice 2

Subtract 47 – 15 mentally.

Break apart the second number to subtract mentally.

Break apart: 15 = 10 + ______

Subtract: 47 – 10 = ______

37 – 5 = ______

47 – 15 = ______

On Your Own

Subtract 59 – 27 mentally.

59 – 27 = ______

Write About It

Could you break apart the second number to find 76 – 29 mentally? Explain.

Objective 8.1: Use mental arithmetic to find the sum or difference of two two-digit numbers.

Add or Subtract Mentally

1. Find the number.

   a. ten more than 35 _______

   b. ten less than 67 _______

   c. twenty more than 38 _______

   d. thirty less than 96 _______

2. Find the sum mentally.

   a. 68 + 12 = _______

   b. 52 + 35 = _______

   c. 14 + 65 = _______

   d. 55 + 15 = _______

3. Find the difference mentally.

   a. 54 − 13 = _______

   b. 60 − 35 = _______

   c. 97 − 45 = _______

   d. 43 − 13 = _______

4. Use mental arithmetic to compute. Which is greater than 56? Circle the letter of the correct answer.

   A 32 + 26 **B** 27 + 20

   C 83 − 50 **D** 94 − 40

5. Dan biked 25 miles one day and 15 miles the next. How far did he bike in all? Circle the letter of the correct answer.

   A 30 miles **B** 40 miles

   C 35 miles **D** 50 miles

6. Dorothy says that it is easy to mentally add two numbers that have 5 in the ones place. Why do you think she says this?

   ____

   ____

7. To add 27 + 17 mentally Kyle used 7 + 7 = 14, 20 + 10 = 30, 30 + 4 = 34. Did he get the right answer? Explain.

   ____

   ____

8. The goal of this game is to create two numbers with the greatest possible sum. Each player makes two 2-digit numbers using each of the digits 1, 2, 3, and 4 once. Joshua makes 32 and 14. Lindsay makes 24 and 31. Who wins the game? Explain.

   ____

   ____

Objective 8.1: Use mental arithmetic to find the sum or difference of two two-digit numbers.

Lesson 8-2 Add and Subtract with Estimation

Words to Know An **estimate** is a number close to the exact answer.

Activity 1

Use the number line to round each number. Then estimate the sum.

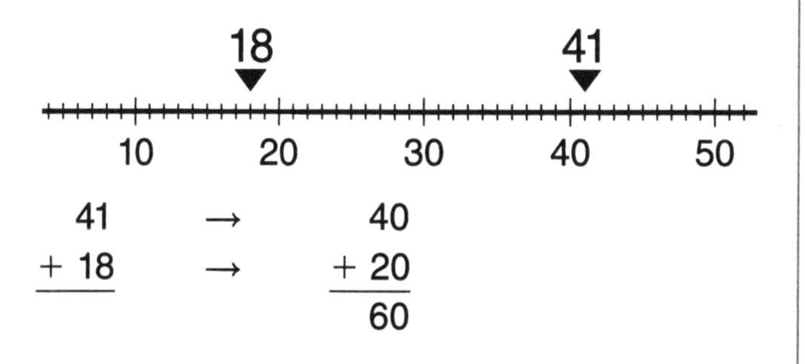


```
  41   →    40
+ 18   →  + 20
            60
```

Practice 1

Use the number line to round each number. Then estimate the sum.

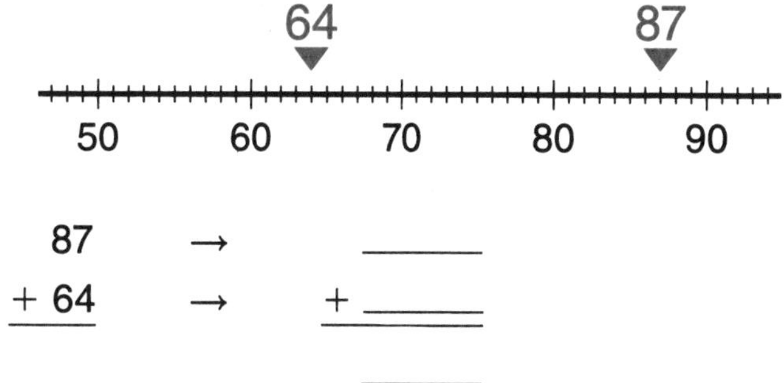


```
  87   →    ____
+ 64   →  + ____
            ____
```

Activity 2

Round to estimate the difference.

```
  55   →    60
– 28   →  – 30
            30
```

Practice 2

Round to estimate the difference.

```
  89   →    ____
– 45   →  – ____
            ____
```

On Your Own

Round to estimate.

a. 73 + 18

Estimate: ______

b. 45 – 29

Estimate: ______

Write About It

Explain how to estimate 49 + 32 by rounding.

Objective 8.2: Estimate sums and differences of 2-digit numbers.

Lesson 8-2 Add and Subtract with Estimation

B
Understand It

Example 1

Use front-end estimation to estimate the sum.

In front end estimation, keep the first digit the same. Change all following digits to 0.

$$\begin{array}{r} 67 \\ +\ 22 \\ \hline \end{array} \quad \begin{array}{c} \rightarrow \\ \rightarrow \end{array} \quad \begin{array}{r} 60 \\ +\ 20 \\ \hline 80 \end{array}$$

Practice 1

Use front-end estimation to estimate the sum.

43 Change the ones digit to 0. ______

+ 28 Change the ones digit to 0. + ______

______

Example 2

Use front-end estimation to estimate the difference.

$$\begin{array}{r} 94 \\ -\ 45 \\ \hline \end{array} \quad \begin{array}{c} \rightarrow \\ \rightarrow \end{array} \quad \begin{array}{r} 90 \\ -\ 40 \\ \hline 50 \end{array}$$

Practice 2

Use front-end estimation to estimate the difference.

76 Change the ones digit to 0. ______

– 28 Change the ones digit to 0. – ______

______

On Your Own

Use front-end estimation.

a. 84 + 17

Estimate: ______

b. 53 – 39

Estimate: ______

Write About It

Explain how to estimate 49 + 32 by front-end estimation.

Objective 8.2: Estimate sums and differences of 2-digit numbers.

Lesson 8-2 Add and Subtract with Estimation

1. Round to estimate.

   a. 67 + 21 Estimate: ______

   b. 53 − 18 Estimate: ______

2. Use front-end estimation to estimate.

   a. 91 + 48 Estimate: ______

   b. 72 − 26 Estimate: ______

3. Round to estimate.

   a. 88 + 63 Estimate: ______

   b. 55 − 19 Estimate: ______

4. Use front-end estimation to estimate.

   a. 67 + 38 Estimate: ______

   b. 54 − 22 Estimate: ______

5. What is the estimate for 68 + 35 when the addends are rounded? Circle the letter of the correct answer.

   A 80 **B** 90

   C 100 **D** 110

6. A pair of shoes costs $48 and a jacket costs $74. About how much do the shoes and jacket cost all together? Use front-end estimation. Show your work.

   The shoes and jacket cost about $______.

7. If each sum is estimated by rounding **and** then by front-end estimation, which sum will have the same estimate from both methods?

   A 12 + 19 **B** 25 + 48

   C 62 + 84 **D** 36 + 91

8. What is the advantage of using rounding? What is the advantage of using front-end estimation?

Objective 8.2: Estimate sums and differences of 2-digit numbers.

Add and Subtract 3-Digit Numbers

Activity 1

Use base ten blocks to solve this problem.
143 + 65 = ?

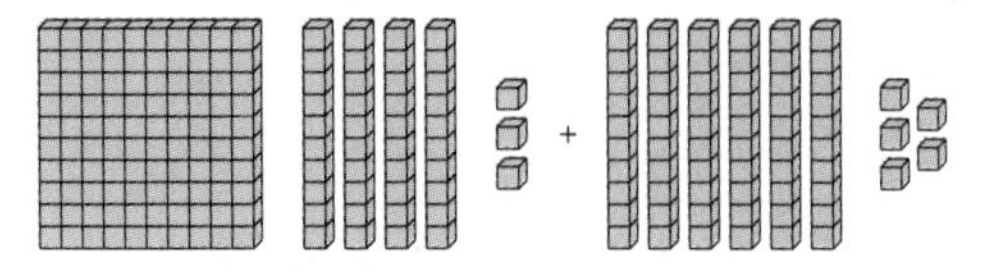

143 + 65 = 208

Model each number.
Combine ones, tens, and then hundreds.
Trade for a bigger block when you can.

Practice 1

Use base ten blocks to solve this problem.

426 + 347 = ?

How many ones in all? _______

Trade _______ ones for _______ tens.

Now there are _______ tens and _______ ones.

There are _______ hundreds.

426 + 347 = _______

Activity 2

Use base ten blocks to solve this problem.

100 − 57 = ?

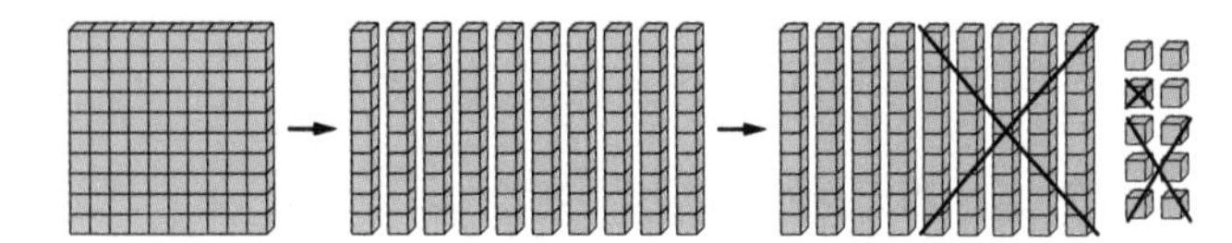

H	T	O
	9	
	~~10~~	10
~~1~~	~~0~~	~~0~~
−	5	7
	4	3

Practice 2

Use base ten blocks to solve this problem.

508 − 214 = ?

H	T	O
5	0	8
−2	1	4

On Your Own

Add. Remember to show any regrouping.

H	T	O
6	2	4
+2	8	5

Write About It

What is the first step when you subtract 430 from 600 using base ten blocks?

Objective 8.3: Find the sum or difference of two whole numbers up to three digits long.

Lesson 8-3 Add and Subtract 3-Digit Numbers

B Understand It

Example 1

Use the number line to help you subtract.
357 − 4 = ?

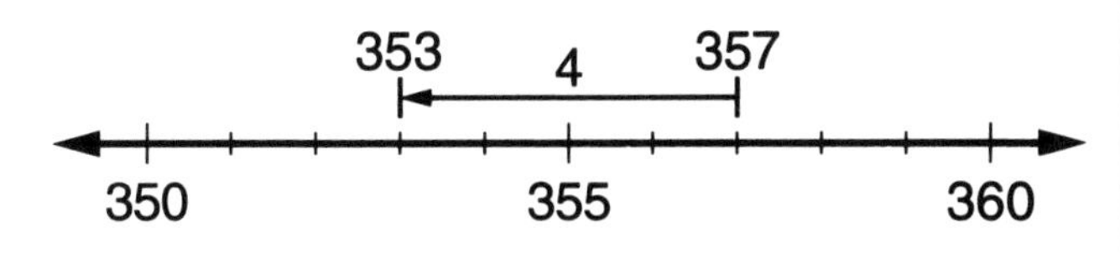

Start at 357.
Move 4 to the left.
You stop at 353.

357 − 4 = 353

Practice 1

Use the number line to help you add.

230 + 50 = ?

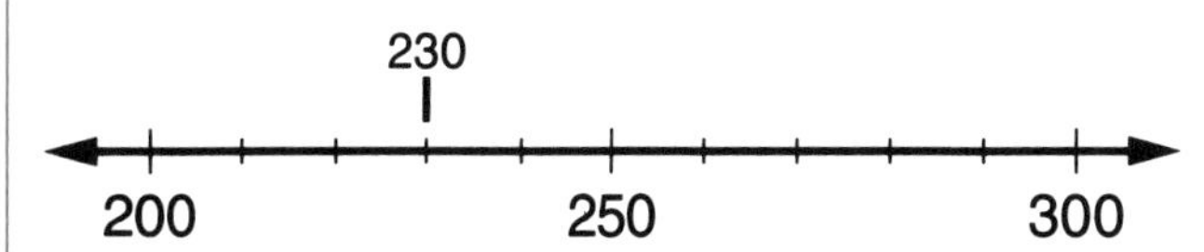

Start at ______.

Move ______ to the right.

You stop at ______. 230 + 50 = ______

Example 2

Spencer has 189 cards. He buys a pack of 25 cards. How many does he have in all?

To find how many in all, add 189 + 25.

$$\begin{array}{r} 189 \\ +\ 25 \\ \hline \end{array} \qquad \begin{array}{r} {}^{1}\ \\ 189 \\ +\ 25 \\ \hline 4 \end{array} \qquad \begin{array}{r} {}^{1\,1}\ \\ 189 \\ +\ 25 \\ \hline 214 \end{array}$$

Spencer has 214 cards.

Practice 2

Room 7 had 145 bracelets to sell at the school history fair. If 97 were left after an hour, how many were sold during that time?

Subtract to find how many were sold.

______ bracelets were sold.

	H	T	O
	1	4	5
−		9	7

On Your Own

Carly has 297 beads. She buys 155 more. How many beads does she have all together?

Carly has ______ beads.

Write About It

Explain how you could use a number line to subtract 200 from 900.

Objective 8.3: Find the sum or difference of two whole numbers up to three digits long.

Lesson 8-3 Add and Subtract 3-Digit Numbers

1. Write the subtraction problem for the picture.

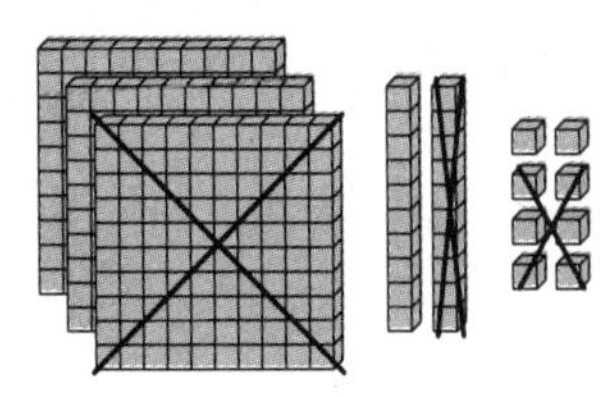

_____ – _____ = _____

2. Use the number line to find the sum or difference.

a. 4 + 135 = _____ **b.** 600 – 30 = _____

3. Add or subtract.

a. 162 + 77 = ____

b. 691 – 59 = ____

c. 483 + 205 = ____

d. 550 – 437 = ____

4. In which subtraction would you need to trade 1 hundred for 10 tens? Circle the letter of the correct answer.

A 983 – 261 **B** 665 – 234

C 792 – 563 **D** 328 – 185

5. Janice says 243 + 369 equals 502. Is her answer correct? Explain why or why not.

____

____

____

6. Is 265 + 43 the same as 200 + 60 + 40 + 5 + 3? Why?

____

____

7. Ali writes a story that is 658 words long. Mike writes a story that is 593 words long. How much longer is Ali's story than Mike's?

____

8. Mr. Wright drove 240 miles in two days. He drove 93 miles on Monday. How far did he drive on Tuesday?

____

Objective 8.3: Find the sum or difference of two whole numbers up to three digits long.

Lesson 8-4 Whole Numbers to 10,000

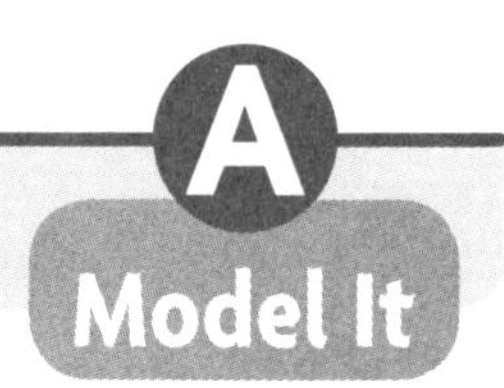

Activity 1

Use base ten blocks to help you add.
343 + 128 = ?

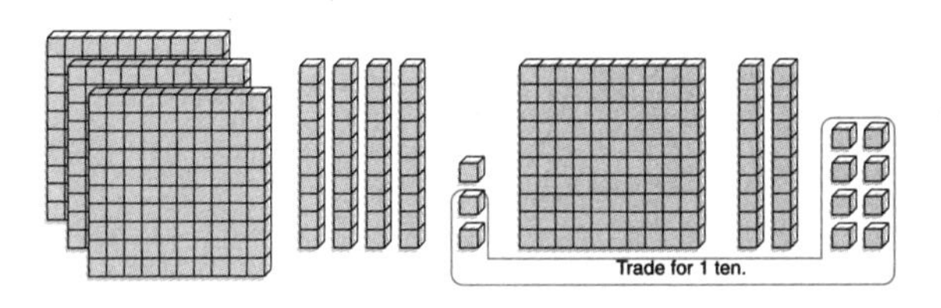

H	T	O
	1	
3	4	3
+1	2	8
4	7	1

Practice 1

Use base ten blocks to help you add.
267 + 261

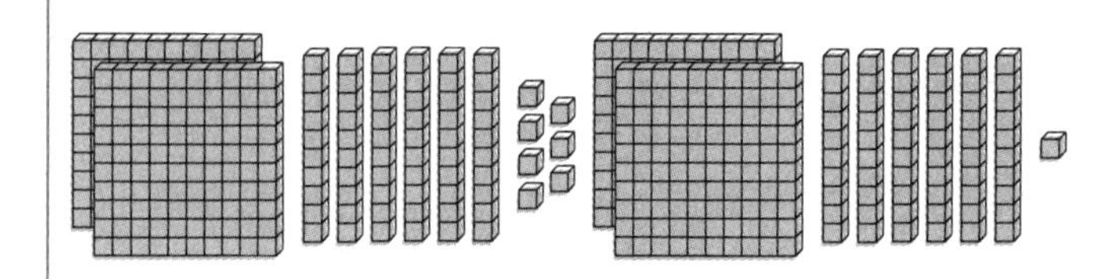

H	T	O
2	6	7
+2	6	1
___	___	___

Activity 2

Use base ten blocks to help you subtract.

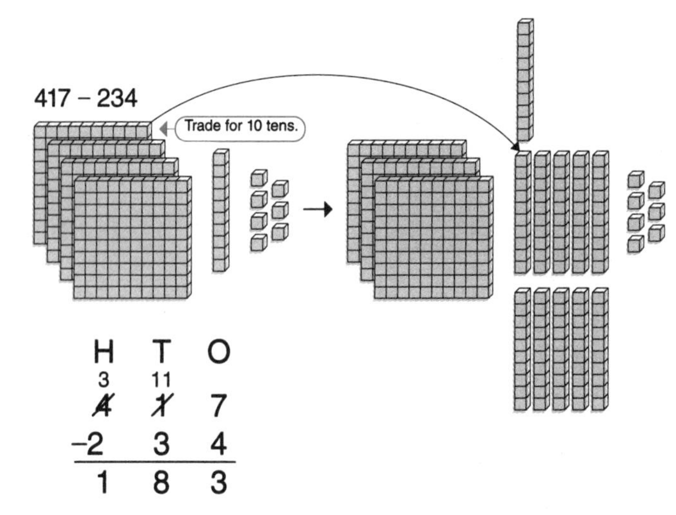

H	T	O
3	11	
~~4~~	~~1~~	7
–2	3	4
1	8	3

Practice 2

Use base ten blocks to help you subtract.

320 – 138

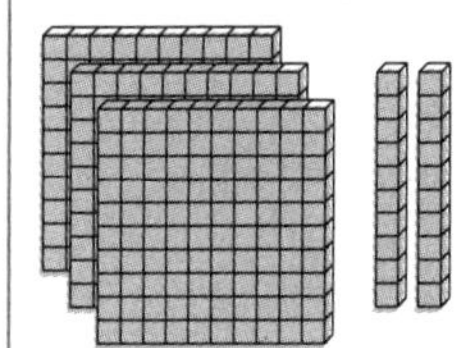

H	T	O
3	2	0
–1	3	8
___	___	___

On Your Own

Subtract.

H	T	O
4	4	1
–2	9	8
___	___	___

Write About It

Charlie added 438 + 163. He wrote the sum as 591. It was marked wrong on his paper. What error did he make?

Objective 8.4: Find the sum or difference of two whole numbers between 0 and 10,000.

Lesson 8-4 **Whole Numbers to 10,000**

B Understand It

Example 1

Add 2,363 and 4,956.

Line up the places.
Add in each place and regroup as needed.

Th	H	T	O
1	1		
2,	3	6	3
+4,	9	5	6
7,	3	1	9

Practice 1

Add 7,631 and 1,293.

Line up the places.
Add in each place and regroup as needed.

Th	H	T	O
7,	6	3	1
+1,	2	9	3
___,	___	___	___

Example 2

Ice Cream Delights had sales of $4,236 the first week and sales of $2,684 the next week. How much greater were the sales the first week than the second week?

Subtract to solve.

Th	H	T	O
	11		
3	~~1~~	13	
~~4~~,	~~2~~	~~3~~	6
−2,	6	8	4
1,	5	5	2

Sales were $1,552 greater the first week.

Practice 2

Mr. Ortega wants to save $3,000 for a family vacation. He has saved $1,562. How much more does he need to save?

Mr. Ortega needs to save ________ more.

On Your Own

Subtract 2,342 − 1,383.

2,342 − 1,383 = ______

Write About It

Why is place value important in addition and subtraction?

Objective 8.4: Find the sum or difference of two whole numbers between 0 and 10,000.

Lesson 8-4 Whole Numbers to 10,000

1. Use the base ten blocks to help you add.
252 + 167 = ?

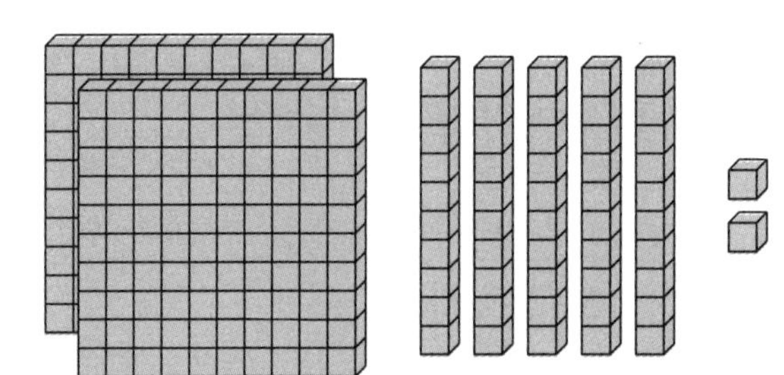

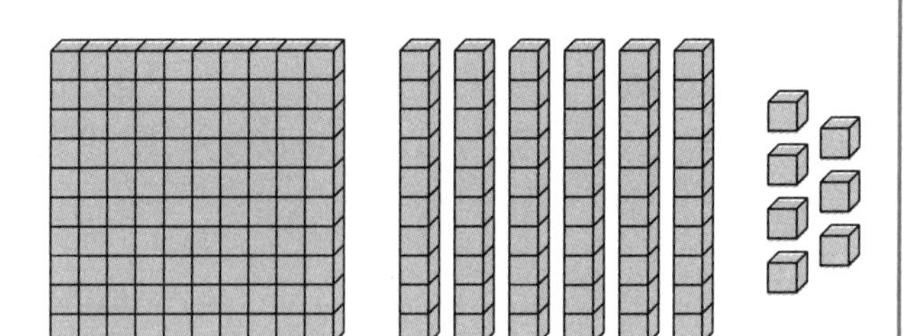

252 + 167 = ______

2. Add.

a. 2,053
+2,884

b. 7,076
+1,359

c. 6,219
+ 312

d. 3,754
+ 46

3. Subtract.

a. 5,890
− 394

b. 8,072
− 1,895

c. 3,555
−1,912

d. 1,684
−1,265

4. What is 2,253 more than 1,592?

A 4,755
B 3,845
C 3,745
D 661

5. The park has an area of 10,072 square feet. The pond in the park has an area of 894 square feet. What is the area of the park without the pond?

____

6. The sum of two numbers is 3,405. If one of the addends is 1,563, what is the other addend? Explain how you know.

Objective 8.4: Find the sum or difference of two whole numbers between 0 and 10,000.

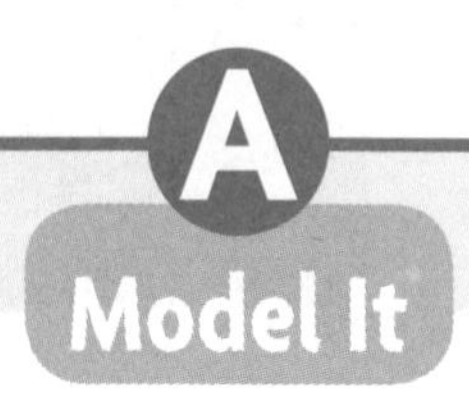

Activity 1

Find 627 + 493.
You can use base ten blocks.

H	T	O
1	1	
6	2	7
+4	9	3
1,1	2	0

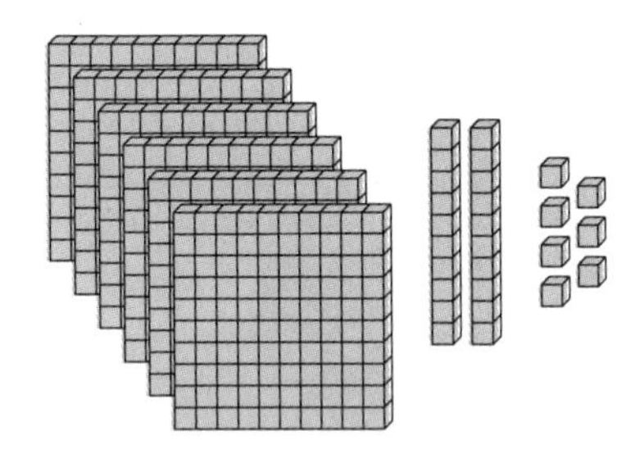

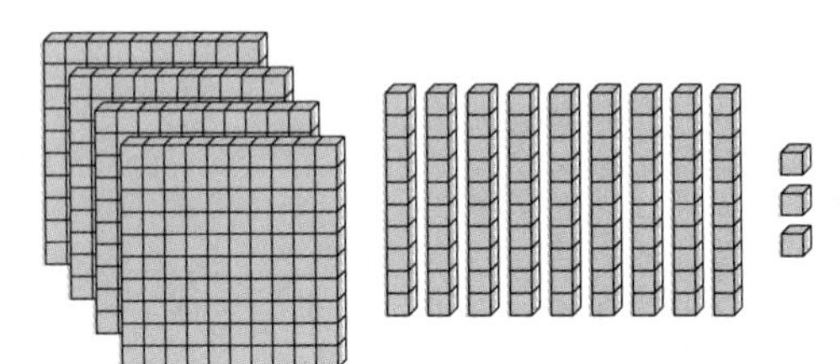

Practice 1

Add 538 + 579. Explain each step.
You can use base ten blocks.

H	T	O
5	3	8
+5	7	9

Add ones. ______________________________

____

Add tens. ______________________________

____

Add hundreds. ______________________________

____

____

Write 1 in the ____________ place.

Activity 2

Round to estimate the sum 4,513 + 83,849. Round each number to the thousands place. Then find the exact sum.

Estimate	Exact
	1 1
5,000	4,513
+84,000	+83,849
89,000	88,362

Practice 2

Round to estimate 101,298 + 71,685. Round each number to the ten thousands place. Then find the exact sum.

Estimate	Exact
100,000	101,298
+ 70,000	+ 71,685

On Your Own

Add 239 + 912.

____

When do you need to trade or carry?

Objective 8.5: Demonstrate an understanding of, and the ability to use, standard algorithms for the addition and subtraction of multidigit numbers.

Multidigit Numbers

Example 1

Subtract: 234 – 151. You can use base ten blocks to help you.

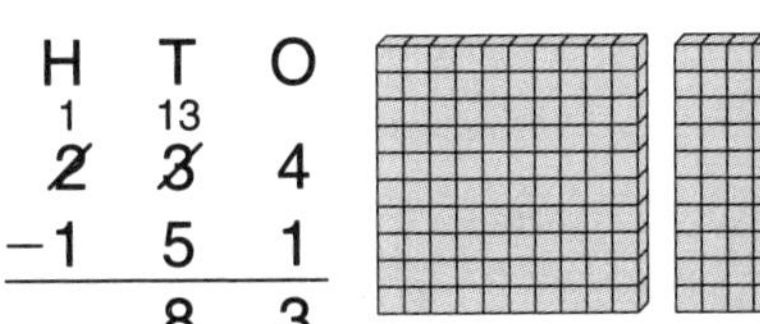

H	T	O
1	13	
~~2~~	~~3~~	4
−1	5	1
	8	3

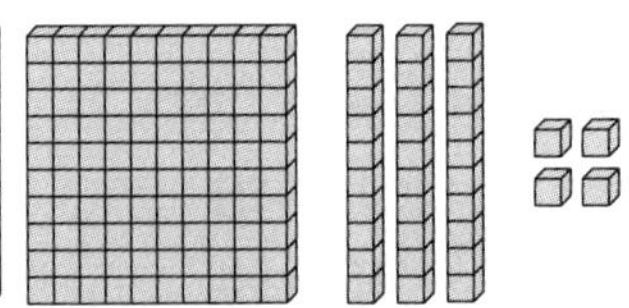

Regroup and subtract.

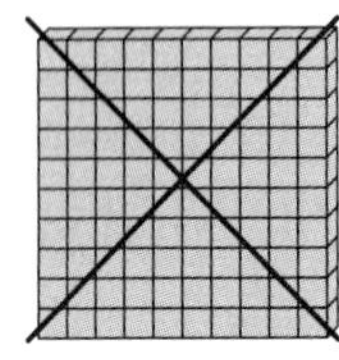
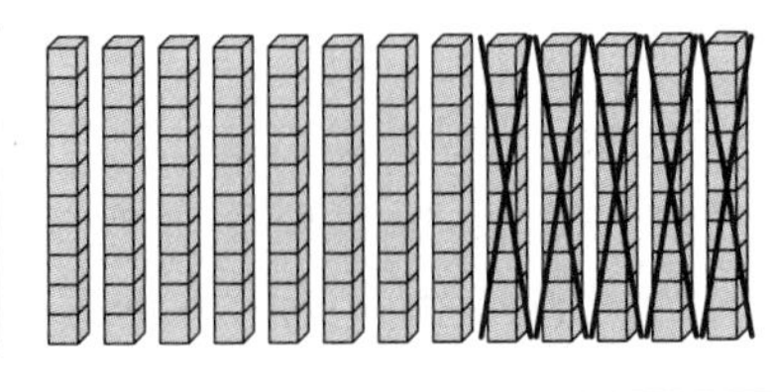

Practice 1

Subtract: 602 – 235.
You can use base ten blocks to help you.

H	T	O
6	0	2
−2	3	5

Example 2

Round to estimate the difference 813,602 – 51,851. Then find the exact difference.

813,602 ↘ 810,000

51,851 ↘ 50,000

Write the problem in the place-value chart.

7 ~~8~~ −	11 ~~1~~ 5	0, 0,	0 0	0 0	0 0
7	6	0,	0	0	0
7 ~~8~~ −	11 ~~1~~ 5	2 ~~3~~, 1,	15 ~~5~~ ~~6~~ 8	10 ~~0~~ 5	2 1
7	6	1,	7	5	1

Practice 2

Round to estimate the difference 752,267 – 17,617. Then find the exact difference.

Write the problem in the place-value chart.

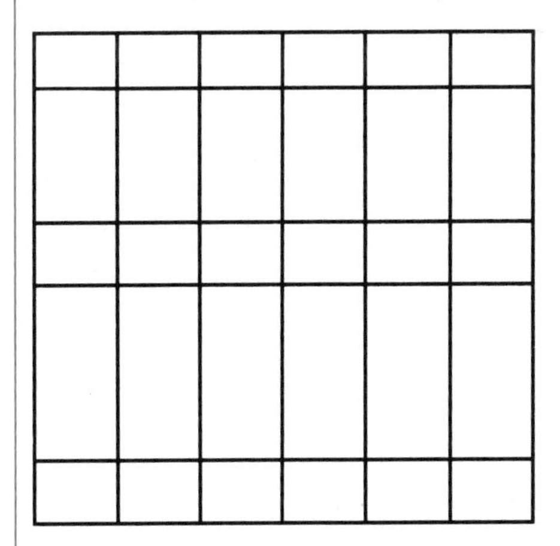

On Your Own

Find 12,524 + 9,362.

Write About It

Give a real-world example to show the idea of trading in subtraction.

Objective 8.5: Demonstrate an understanding of, and the ability to use, standard algorithms for the addition and subtraction of multidigit numbers.

Multidigit Numbers

1. Subtract 425 – 129. Show your work.

2. Solve.

a. $\begin{array}{r} 4{,}275 \\ +\ 5{,}349 \\ \hline \end{array}$

b. $\begin{array}{r} 906 \\ +\ 1{,}508 \\ \hline \end{array}$

c. $\begin{array}{r} 21{,}916 \\ -\ 10{,}634 \\ \hline \end{array}$

d. $\begin{array}{r} 23{,}070 \\ -\ 13{,}370 \\ \hline \end{array}$

3. Estimate. Round to the nearest ten thousand.

a. 20,111 + 16,479 = about ________

b. 691,791 – 85,942 = about ________

4. The areas of California and Alaska are 163,696 sq mi and 591,004 sq mi. What is the difference of their areas?

____

5. One year Mrs. Rodriquez earned $32,565. She paid $5,629 in taxes. What was her income after taxes? Circle the letter of the correct answer.

A $26,936 **B** $32,003

C $33,144 **D** $37,194

6. What is 12,500 less than 81,056? Circle the letter of the correct answer.

A 61,456 **B** 68,556

C 71,556 **D** 93,556

7. When adding two numbers do you ever trade 20 ones for 2 tens? Explain why or why not.

____

____

____

____

8. Mount Whitney in California is 14,494 feet high. Black Mountain in Kentucky is 4,139 feet high. About how much higher is Mount Whitney than Black Mountain? Explain.

Objective 8.5: Demonstrate an understanding of, and the ability to use, standard algorithms for the addition and subtraction of multidigit numbers.

Topic 8 Add or Subtract Multidigit Numbers

Topic Summary

Choose the correct answer. Explain how you decided.

1. Tonya has a collection of coins from other countries. If she has 29 coins from Mexico and 43 coins from Canada, about how many coins does she have in all?

A 62 **B** 14 **C** 72 **D** 70

2. At a basketball game, there were 12,496 fans of the home team and 8,757 fans of the visiting team. How many more fans of the home team were at the basketball game?

A 21,253

B 3,000

C 3,749

D 3,739

Objective: Review addition and subtraction of multidigit numbers.

Add or Subtract Multidigit Numbers

Mixed Review

1. Find the sum.

   a. $16 + 23 =$ ______

   b. $79 + 11 =$ ______

   c. $34 + 19 =$ ______

Volume 2, Lesson 7-2

2. Write the number three thousand, seven hundred eighty-nine in standard form.

____

Volume 1, Lesson 2-3

3. Use the cubes to model $12 - 3$.

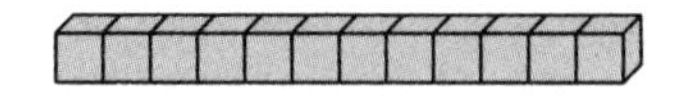

$12 - 3 =$ ______

Volume 2, Lesson 4-6

4. Solve.

$$\begin{array}{r} 256 \\ +\ 39 \\ \hline \end{array}$$

Volume 3, Lesson 8-3

5. Kayla estimated $43 + 17$ to $50 + 20$.

   a. What error did she make?

   ____

   ____

   ____

   b. Estimate and solve. ______________

Volume 3, Lesson 8-2

6. Which of the following is a fact in the family 5, 6, and 30? Circle the letter of the correct answer.

   A $5 + 6 = 30$

   B $30 \div 5 = 6$

   C $30 \times 6 = 5$

   D $30 - 5 = 6$

Volume 2, Lesson 6-5

Objective: Maintain concepts and skills.

Words to Know/Glossary

estimate — A number close to the actual answer.

R

regroup — Exchange amounts of equal value to rename a number.

Word Bank

Word	My Definition	My Notes

Word	My Definition	My Notes

Index

A

M

R

S

T